WORLD OF WONDER

Published by Creative Education
123 South Broad Street
Mankato, Minnesota 56001

Creative Education is an imprint of
The Creative Company.

Art direction by Rita Marshall
Design by The Design Lab
Photographs by Bradley Ireland Productions, Richard Goff,
The Image Finders (Eric R. Berndt, Michael Evans, Mark &
Sue Werner), JLM Visuals (Richard P. Jacobs, Breck P.
Kent), Robert McCaw, Premiere Stock (James Prout), James
P. Rowan, Tom Stack & Associates (Joe McDonald, Milton
Rand), Frank Staub, Unicorn Stock Photos (Robert W. Ginn)

Library of Congress Cataloging-in-Publication Data

Staub, Frank J.
The food chain / by Frank Staub.
p. cm. — (World of wonder)
Summary: Describes the interaction of predators, prey,
plants, and non-living elements that make up the food chain,
and touches on what happens to the food chain when the
balance of nature is upset.
ISBN 1-58341-269-7
1. Food chains (Ecology)—Juvenile literature. [1. Food Chains
(Ecology). 2. Ecology.] I. Title. II. World of Wonder
(Mankato, Minn.).

QH541.14.S725   2003
577'.16—dc21   2002035145

First Edition

9 8 7 6 5 4 3 2 1

cover & page 1: a chipmunk eating a raspberry
page 2: a puffin holding fish
page 3: a bighorn sheep feeding

Creative Education presents

W O R L D   O F   W O N D E R

THE FOOD CHAIN

BY   FRANK   STAUB

Grass grows in the sun ❧ A prairie dog gets fat munching on the grass 🐀 A badger pounces on the prairie dog ❄ This flow of food from organism to hungry organism is called a food chain. There are fascinating food chains in all habitats, from jungles and oceans to mountains and plains.

🌐

MOST FOOD CHAINS on land begin with green plants. Animals eat the plants. The plant-eaters may then be eaten by other animals, which may be eaten by still other animals. It seems simple. But each living link in a food chain forms just one strand in a complex web of life.

A prairie dog feeds on blades of green grass

FOOD PRODUCERS

Nearly every living thing on Earth, from the smallest bug to the largest tree, depends on sunlight for food. *Photosynthesis* is a Greek word that means "making with light." It's the process by which green plants, **algae**, and certain other organisms make food by combining the **energy** in light with water and a gas called carbon dioxide.

NATURE NOTE: *Every inch of a giant sequoia tree, the largest living thing on Earth, is made from food produced by photosynthesis in its leaves.*

Organisms that make their own food are called producers. Grasses are the main producers on North American prairies. **Phytoplankton** are the main producers in oceans and lakes. Producers use the energy in their food for everyday tasks such as fighting disease and making flowers. They also turn food into the materials of which their bodies are made.

NATURE NOTE: *Chlorophyll, a green substance that traps the energy of light during photosynthesis, gives the leaves of most plants a green color.*

Many producers, such as orange trees, use the energy in their food to yield fruit.

✳ Producers form the base, or bottom, of every food chain. For organisms higher in the food chain, producers are the primary source of energy and **nutrients**. They are a food chain's first link.

NATURE NOTE: *Dinosaurs may have died out when an asteroid hit Earth and clouded the sky with dust. Sunlight couldn't reach the plants, the plants died, and the dinosaurs starved.*

Photosynthesis produces all of the fruits we eat

ANIMAL DIETS

Animals can't make food, so they have to eat other organisms. For example, a squirrel can't make the acorns it eats. But photosynthesis in oak tree leaves provides the energy and materials the trees need to produce acorns.

Squirrels eat plants and plant parts. They're herbivores, or plant-eaters. Elk, katydids,

NATURE NOTE: *Pitcher plants are both producers and carnivores. They perform photosynthesis and eat insects trapped in the water-filled tubes formed by their leaves.*

Squirrels feed on trees' acorns and pine cones

and chuckwalla lizards are other examples of herbivores that live on land. Ocean herbivores include green sea turtles and many **species** of fish and **zooplankton**.

❧ Mountain lions and wolves don't eat plants. They're carnivores, or meat-eaters. Cats, frogs, and tarantulas are other examples of land carnivores. Sharks, sea lions, and starfish are carnivores of the sea. These and other **predators** get all their nutrients from their **prey**.

Wolves dine on other animals such as squirrels

Raccoons feed on both plants and animals. Animals that have such mixed diets are called omnivores. Box turtles, bears, sandhill cranes, and people are all examples of omnivores.

NATURE NOTE: *A predator at the top of its food chain—such as a jaguar—is usually big, powerful, and fierce, with no enemies other than man.*

A raccoon's diet includes fish, mice, and fruit

NATURE NOTE: *Most herbivores eat only certain kinds of plants. But mountain goats eat any plant they can find because there are so few plants in their cold, rocky habitats.*

BREAKING IT DOWN

When a great blue heron gulps down a fish, it eventually gets rid of the bones and other unusable fish parts by way of its body waste. This waste becomes food for **bacteria**, **fungi**, and other

Herons and other seabirds swallow fish whole

decomposer organisms. Decomposers also eat dead organisms, as well as cast-off parts such as leaves and feathers. As they eat, the decomposers break their lifeless food down into simple substances that living plants absorb as nutrients.

❧ On a forest floor, leaves, twigs, and other plant parts pile up to form a layer called litter. While decomposers break the

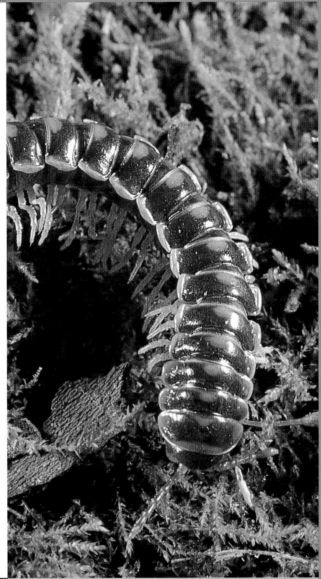

NATURE NOTE: *Mushrooms grow from decomposer fungi in the soil. When slugs eat mushrooms, they become part of a food chain that is supported by waste and dead organisms.*

Millipedes help keep the ground clear of waste

litter down into nutrients, slugs, millipedes, earthworms, and other small creatures serve as natural cleaning crews by eating the decaying litter and the decomposers.

🐸 Countless sea creatures depend on dead plant material from mangrove swamps and salt marshes along seacoasts. As salt marsh grasses and mangrove tree leaves die and fall into the

NATURE NOTE: *Unlike green plants, snow plants in Oregon and California lack chlorophyll for photosynthesis. They absorb all their food from materials decomposing in the soil.*

water, they provide a feast for decomposers. The plant pieces and decomposers mix to form a kind of high-energy, nutrient soup that feeds worms, snails, and other little sea creatures. These are the first links in food chains containing many of the world's fish and birds.

🕷 Animals that eat waste or dead animals are called scavengers. Vultures are well-known scavengers. Fishers, eagles, and many other predators will become part-time scavengers if they come across a free meal in the form of a dead animal.

Salt marshes support many ocean food chains

NATURE NOTE: *Arctic ground squirrels on the tundra make up about 90 percent of the golden eagle's diet and 50 percent of the red fox's diet.*

LIVING WEBS

Food chains are often intertwined. A squirrel may fall prey to a wolf, coyote, mountain lion, or fox. These predators also eat rabbits, mice, and birds. Foxes eat berries during the summer. Birds, rabbits, and mice eat berries, too. Raccoons will eat insects. So will foxes. Wolves and mountain lions prefer deer, but sometimes they become scavengers. This complex network of food chains, called a food web, is typical of most natural communities.

Some food webs are more complicated than others. On the African plains, grass, trees, and bushes produce food for zebras, birds, antelope, and hundreds of other plant-eating species. These herbivores fall prey to lions, leopards, cheetahs, and hyenas. Dung beetles eat the animals' waste and are preyed upon by honey badgers and many birds. Vultures, hyenas, and jackals scavenge the dead. If one species disappears, predators and scavengers of the missing link have other creatures to choose from.

NATURE NOTE: *Rattlesnakes eat packrats and other small animals. The energy lost between herbivores and carnivores is the reason there aren't more rattlesnakes in desert food chains.*

Mice are an abundant food source for rattlesnakes

For this reason, African plains food chains are known as stable food chains.

✳ The African plains' sunny, warm weather is good for plants. But on the **tundra**, summers are so short that plants have little time to grow. Few plants, and fewer animals, survive in such cold places. So tundra food chains are short. They're also unstable. For example, arctic

NATURE NOTE: *Most herbivores, such as moose, obtain only about 10 percent of the energy that their plant food captured from the sun by photosynthesis.*

ground squirrels are one of the few small tundra animals. But they're a major food source for foxes, bears, wolves, and golden eagles. If disease were to kill a lot of arctic ground squirrels, some of these predators might starve.

NATURE NOTE: *Predators with big appetites may travel great distances to find enough prey. A mountain lion may hunt for food across 100 square miles (256 sq km).*

Bears have limited food options on the tundra

ENERGY LOST

Crayfish and other small pond animals eat water plants. These creatures are prey for larger animals such as soft-shelled turtles, which are prey for alligators. This food chain has four levels. Some food chains have five. But few have more. The reason is energy. As energy moves from plants to small animals to larger animals, some gets used for activities such as nesting and hunting, and some passes out of the animals in their waste. By the time the energy reaches a large predator such as an alligator, it's just a fraction of what the plants first captured during photosynthesis.

All of the producers in a community of plants and animals contain more energy than all of the herbivores. Likewise, the herbivores contain more energy than all of the carnivores. This explains why there's

NATURE NOTE: *Raising cattle requires more energy than raising corn. If people ate more plants and less meat, food would cost less and there would be more to go around.*

23   *Plentiful plants contribute to high deer populations*

usually more producers than herbivores and more herbivores than carnivores in a food chain.

❧ Big animals need a lot of energy. So bison, moose, and manatees eat plants at the bottom of the food chain. This makes sense because that's where most of the food is. If these giants ate other animals, they might not find enough prey to satisfy their huge appetites. Much of the energy captured by photosynthesis would get used up before it reached them.

Bison may eat 30 pounds (14 kg) of food a day

TOP-HEAVY CHAINS

Some food chains seem to be top-heavy because the herbivores outnumber the producers. For instance, in some habitats, thousands of insects may feed on just one tree. This food chain works because the tree contains more material and weighs more than all of the insects feeding on it combined.

🌿 The weights of organisms can make some food chains seem top-heavy.

A single tree can support a huge colony of ants

Phytoplankton in a pond usually outnumber the zooplankton that eat them. Yet all of the zooplankton together may weigh more than all of the phytoplankton. How is this possible? The food chain works because phytoplankton reproduce quickly. As some get eaten, others are born.

Weights and numbers may not always offer an accurate picture of how the different levels in a food chain are related. But energy does. Tree leaves always capture more

energy from the sun than caterpillars get from eating the leaves. And zooplankton will never get as much energy from eating phytoplankton as the phytoplankton obtain from the sun during photosynthesis.

NATURE NOTE: *Most of the ocean fish caught by commercial and recreational fishermen are part of food chains that depend on salt marshes and mangrove swamps.*

POISON BUILDUP

In the 1960s, a poison called DDT was sprayed in parts of the United States to kill insects that destroyed crops and spread disease. Rainwater washed much of the DDT into streams and lakes, where phytoplankton absorbed it. By way of the food chain, the DDT got into zooplankton that ate the phytoplankton, the little fish that ate the zooplankton, and the big fish that ate the little fish.

☼ The poison ended up in bald eagles, pelicans, and other fish-eating birds. Each fish a bird ate contained most of the DDT from everything below it in the food chain. The high levels of DDT in their bodies caused the birds to produce eggs with very thin shells. All too often, the eggs broke before hatching. Partly because of this, the bald eagle population outside Alaska dropped to just 3,000 birds. Since DDT was banned in 1972, bald eagle numbers have risen to about

NATURE NOTE: *Scientists think that, through food chains, white pelicans in southern California may be taking in large quantities of DDT used in Mexico.*

70,000. But a century earlier, there were at least 250,000.

🕷 Today, the use of DDT is against the law in many places. But some countries still spray the poison to kill mosquitoes. And unfortunately, lead, mercury, and other dangerous chemicals from factories, farms, and dumps continue to poison some of the world's food chains.

NATURE NOTE: *In the 1950s, mercury dumped into the ocean by a Japanese factory entered food chains and caused birth defects and death in some people who ate seafood.*

Pesticides protect crops but can poison food chains

NATURE'S BALANCE

Energy and nutrients continually move through Earth's many food chains. Plants reuse nutrients after decomposers break down waste and dead organisms. But energy passes through a food chain only once. Then it's gone. Fortunately, photosynthesis is always capturing more energy from the sun to keep the food chains going.

❄ The flow of food and energy from one organism to another helps living things thrive in the challenging world around them. The world's food chains are valuable reminders of how intricately the lives of various creatures are intertwined. As humans make changes that affect the environment, it's important to remember and respect these connections. In doing so, we can help ensure the future health and beauty of this amazing world, this world of wonder.

NATURE NOTE: *America's deer population is very high, in large part because ranchers and hunters have killed most of the wolves, mountain lions, and other predators that eat deer.*

WORDS TO KNOW

Algae *are living things that have chlorophyll for photosynthesis, but no roots, stems, or leaves.*

Bacteria *are tiny organisms that break down waste and dead organisms; some of them can cause disease.*

Energy *is the ability to do work; all living things need energy to survive.*

Fungi *are a group of organisms that can't perform photosynthesis and include molds, mildews, yeast, and mushrooms.*

Substances that a living thing needs to live and grow are called **nutrients***.*

Phytoplankton *are tiny plants that float in the ocean and other bodies of water.*

Predators *are animals that kill and eat other animals.*

Animals that are eaten by predators are called **prey***.*

A **species** *is a group of living things that can successfully breed with each other.*

A large, treeless expanse of land found in the far North or mountains is called a **tundra***.*

Zooplankton *are tiny animals that float in the ocean and other bodies of water.*

INDEX